11/13

# INTERNSHIP
# & VOLUNTEER
# OPPORTUNITIES

## For People Who Love to Build Things

Laura La Bella

ROSEN
PUBLISHING®

New York

Published in 2013 by The Rosen Publishing Group, Inc.
29 East 21st Street, New York, NY 10010

First Edition

**Library of Congress Cataloging-in-Publication Data**

La Bella, Laura.
Internship & volunteer opportunities for people who love to build things/
Laura La Bella.
    pages cm—(A foot in the door)
Includes bibliographical references and index.
ISBN 978-1-4488-8299-1 (library binding)
1. Building trades—Vocational guidance. 2. Construction industry—
Vocational guidance. 3. Volunteer workers in engineering. 4. Volunteer
workers in architecture. I. Title. II. Title: Internship and volunteer opportunities for people who love to build things.
TH159.L3 2013
624.023—dc23

                                  2012014429

*Manufactured in the United States of America*

CPSIA Compliance Information: Batch #W13YA: For further information, contact Rosen Publishing, New York, New York, at 1-800-237-9932.

# Contents

A senior engineer inspects a turbine while pointing out what to look for to an intern. In construction and architecture, engineers are members of a larger team that includes carpenters, construction workers, woodworkers, and others who all contribute to the craft of building.

# Introduction

A2,723 feet (830 meters), the Burj Khalifa skyscraper in the Middle Eastern city of Dubai is the world's tallest human-made structure. Made of concrete, steel, aluminum, and glass, the structure boasts a number of records, including the world's highest elevator, the highest outdoor observation deck, the most floors of any building worldwide, and the world's highest occupied floor.

Skidmore, Owings and Merrill, from Chicago, Illinois, were behind the tower's architecture and engineering. As the firm crafted a design, architects needed to take into consideration atmospheric conditions at such a height, which influenced their design, as well as cultural and historical elements particular to the region. To support the unprecedented height of the building, engineers developed a new structural system called the buttressed core, which consists of a hexagonal core reinforced by three buttresses that form a "Y" shape. This structural system enables the building to support itself laterally and keeps it from twisting in high winds.

A building as significant and important as the Burj Khalifa is created by a team of professionals. Architects and designers, engineers, construction workers, and carpenters each have a role in the success of a project of this magnitude. While architects and engineers create the initial design of the building and its structure, construction workers implement the design and oversee its physical creation. After the building has been built, carpenters create framing for interiors walls, install drywall, doorways, and other fittings.

For people who like to build things—such as woodworkers, boat builders, carpenters, architects, construction workers, and engineers—getting started in the field takes initiative. These professionals, from the chief architect to the carpenter installing the flooring, all started at the bottom and worked their way up. Through college courses and degrees, apprenticeships, internships, and volunteer opportunities, these professionals gained the experience they needed to land jobs in their respective careers. Internships and volunteer opportunities are the best way to learn about a career field, gain some experience, and make the connections needed to start a career in the building industry.

# Chapter One

# WHY INTERN OR VOLUNTEER?

Internships and volunteer opportunities are the single most important way to gain experience in a career field. It is well documented that college students with internship experience on their résumés get more job offers and better salaries than those with no experience. Internships and volunteer experience expose students to the day-to-day workings of a career field, offer entry-level training, and allow students to apply the knowledge they have gained in the classroom to a real-world setting. These opportunities also help narrow one's career objectives and give students connections to professionals in the field.

## What Is an Intern?

An internship offers on-the-job training. Interns are usually college students looking to gain experience in a career field before they graduate. Interns can also be high school students interested in learning more about a career field before choosing a major in college.

Internships offer you the opportunity to learn in a hands-on environment, where professionals can share their knowledge with you and help you enhance your skills.

The length of an internship can vary depending on the company and the opportunity, but usually it is a short-term assignment lasting a few weeks to a few months. Interns are assigned to a supervisor who oversees their work experience. The supervisor assigns tasks, assists in training the intern, and evaluates his or her performance. In both high school and college, internships can be completed for credit. In this case, a supervisor ensures that the required learning objectives are taking place and submits proof of this.

## Benefits of an Internship

For both the employer and the intern, the benefits of an internship are extensive. Employers are willing to hire interns with little or no experience. Interns are often extra and cost-effective staff that allows a company to take on more projects or accomplish tasks it could not otherwise complete without additional hiring and expense. For interns, working in their desired career field teaches them valuable lessons that can only be learned on the job. Whether paid or unpaid, the benefits of doing an internship reach far beyond a paycheck. These benefits include:

- **Gaining valuable work experience:** While classroom learning often teaches the principles, practices, and required skills of a career field, it's on-the-job training that gives students the opportunity to apply this theoretical learning in a real-world setting. An internship provides the hands-on work experience a student can't get in a classroom. Reading about how to build a guitar is a very different experience from working in the factory and learning how to shape wood so that its density and thickness creates the best acoustic sound.
- **Giving yourself an edge in the job market:** Employers look at more than just your schoolwork or college major when they review your résumé. On-the-job experience shows them you are familiar with their business or industry, you have some entry-level skills, and you can hit the ground running if hired. Many employers prefer applicants who have done an internship or have relevant work experience. In many of the more competitive job markets, it is essential to set yourself apart from others competing for the same job.
- **Boosting your résumé:** Listing an internship or volunteer experience boosts your résumé and makes you more attractive to employers. In a competitive job market, when there are hundreds of applicants for every one job, every piece of experience that sets you apart is key.

# Make the Most of Your Internship

1. **Take the initiative:** Ask for opportunities to perform certain tasks or volunteer when a task needs to be assigned to someone.
2. **Meet as many people as possible:** Mingle with staff in other departments and ask what they do and how they got started in their careers. Shadow employees whose job is related to the one you are interested in. You might find out about a job you never knew existed, or you may get a privileged inside look at how your industry really works.
3. **Attend professional events:** If your internship offers networking opportunities or educational seminars, take advantage of them. The whole reason you are doing an internship is to broaden your view of an industry and gain experience.
4. **Work on a variety of projects:** This gives you a lot of different types of experience, which will reap benefits once you start applying for full-time work in your career field.
5. **Ask for feedback on your performance:** Take the time to ask how you are doing. Constructive feedback can help you learn from your mistakes. Positive written evaluations can also be shown to prospective employers.
6. **Ask questions:** An internship is your opportunity to learn as much as you can about a career field or industry. Ask your supervisor or coworkers about their educational backgrounds, if they would do anything differently, how the job has changed since they started, how it is likely to change in the future, how the company works, and how promotions and raises are awarded. This is the time for you to learn not only about your career field, but also about the industry as a whole.

- **Proving your skills:** You're not the only one who benefits from your internship. Employers view interns as prospective employees. It's one way for a company to really see what type of employee you will be, if you are capable of the work, if you will work hard, take initiative, and contribute to the company. Many companies use their internship programs as a way to scout talent for future jobs.

- **Determining if you're on the right career path:** An internship is great experience if you are certain about which career path you want to pursue. But, believe it or not, it is also the best way to determine if a career is wrong for you. If you think that a video game designer has nothing but fun playing games all day long, for example, you might be surprised to learn that most game designers and developers spend long hours sitting at a computer, programming code and troubleshooting design flaws. The only true way to learn what a career is like is to intern. Since internships are short-term, you can experience a career field without fully committing to it. This experience can give you the inside knowledge to decide if a career is, or isn't, the right choice for you.

- **Network with professionals:** Internships are a great way to meet people in your potential field. Networking is one of the most important ways to land a job. Internships give you the opportunity to meet people who do what you want to do, but they also give you a chance to show off your capabilities, work ethic, and personal and professional qualities. Later on, when jobs open,

Internships give students the chance to apply the skills and knowledge they learn in the classroom to real-world problems in industry.

you are more likely to be thought of as a great candidate because you've had the chance to prove yourself. Industry professionals can also help you connect to other professionals in the field who are in a position to hire, and they can act as references who can speak to your on-the-job abilities.

- **Apply classroom knowledge:** Classroom learning is essential to understanding concepts and theories. Without experience applying what you've learned, however, you're only getting half of the education you need. An internship gives you the chance to apply the knowledge and skills you have acquired in the classroom to real-world problems.
- **Gain confidence and maturity:** Experience builds confidence. An internship will give you the chance to gain experience that you can talk about on job interviews later. You'll also feel more confident when applying for jobs, knowing that your internships will make you a more marketable, experienced, knowledgeable, and impressive candidate than those with little to no experience. As a bonus, when you are interviewing for a job and you are asked if you know how to do a particular task, you can answer positively and discuss exactly what you learned and accomplished during your internship.

## Great Experience Regardless of a Paycheck

Whether an internship is paid or unpaid, it's the experience that counts. Other than reimbursement, there's no qualitative difference between paid or unpaid internships, and one doesn't give you better experience than the other. There are particular industries that are more likely to offer paid internships.

# Volunteer Opportunities
## for People Who Like to Build Things

These are just a few of the many volunteer opportunities that exist for people who like to build things. Be creative, and don't be shy about offering your time and energy to any individuals or organizations that are performing work that appeals to you. You may be surprised at how happy they are to have your help.

- Set-building for your school's drama department.
- IT, networking, computer troubleshooting/maintenance for your school's computer lab, server, and network.
- Repairs and maintenance in your school's AV department.
- Summer maintenance, repair, and building projects at your school.
- Help with your school's landscape and sports field design and maintenance.
- Maintenance, repair, construction/addition, and furniture-building projects at your church, community center, or shipyard/marina.
- Serve as after-school helper, assistant, or "go-fer" to your wood and metal shop instructors.
- Join a work crew for charitable or public construction, repair, and housing projects.
- Join a public works or city parks work crew.
- Volunteer your services to a local, state, or national park service.
- Offer your volunteer services as a handyman, IT troubleshooter, electronics repairman, or carpenter to family, friends, neighbors, school, and community.

Paid internships are most common in the fields of medicine, architecture, engineering, law, business, and technology. Stipends range from $20 to $40 per hour to several thousands of dollars per month for students who are in college. Paid internships often provide enough money to cover living expenses, and many students are able to save some of their salary. Some companies even pay for housing for their interns. Unpaid internships are common in the fields of education, communications, journalism, real estate, and fashion/clothing design.

Even without a paycheck, unpaid internships pay off. In addition to all the personal and professional benefits of an internship outlined above, you will also gain valuable experience, exposure to workplace conduct/expectations, and future references, all without being tied to a long-term contract and commitment.

## Internships vs. Co-ops

Internships are short stints of work, usually lasting from a few weeks to a few months, and devoted to on-the-job learning. A co-op, short for cooperative education, offers a longer work period, typically a semester, summer, or quarter. Co-ops offer you a different experience than an internship. Because the nature of a co-op is to spend a longer period of time working for a company, employers tend to give co-ops more meaningful work assignments and larger responsibilities within a project. Since co-op work periods are longer, students can spend

On-the-job experience is the surest way to learn within your chosen career field. It's also a great way to make contacts and get your foot in the door of a company where you may want to work permanently.

more time on a project, maximizing their involvement, gaining valuable experience, and even being able to witness firsthand the outcome of their involvement in a project.

Many co-ops are also paid, which can help students offset college expenses or allow them to save for life after graduation. The only downside to co-ops is that many occur during the school year. Students will find themselves having to extend their time at college to accommodate the break from coursework that they take for the co-op.

Many universities have extensive co-op programs and assist students in identifying opportunities, arranging leave from school, and planning students' coursework to accommodate their absence.

# Chapter Two

# CAREER EXPLORATION:
## WHAT DO YOU WANT TO BE WHEN YOU GROW UP?

Deciding what career path you want to pursue is exciting, but it can also be overwhelming. With so many interesting and rewarding career choices to choose from, it's hard to know which direction best fits your interests. Finding the path that meets your career aspirations takes time and requires some thought. Understanding what you like to do, what abilities you have and skills you can acquire, and the type of job you want to have will all help you zero in on a career that is right for you.

Exploring a career field is not the same as searching for a job. A job search is often for a position that fulfills a short-term need, such as a summer job. Career exploration is a longer process that includes learning all about an industry, understanding your interests, and pursuing the appropriate education and training. Career exploration is an investment of your time and education toward building long-term employment and professional development within a field.

# How to Explore Careers

Exploring careers can be a fun activity. It gives you a chance to learn about yourself, your likes and dislikes, the types of jobs that best fit your personality and abilities, and where your true interests lie. Self-assessment and aptitude tests often are the first type of activity a career counselor will offer you in order to identify your strengths and interests. But there are things you can do on your own to begin exploring careers you find interesting.

- **Familiarize yourself with job descriptions:** So you know you like to build things, but what does that mean exactly? An architect, for example, designs the interiors and exteriors of buildings. A carpenter might make cabinets or furniture. Will you work at a drafting table or computer most days or be on-site, overseeing construction or doing physical labor? The responsibilities for each type of job vary greatly. Job descriptions will give you insight into the types of tasks you will perform on a day-to-day basis.

- **Learn what education/training is needed to enter the field:** Like any career, you'll need specialized education and experience before

Job fairs are an excellent way to learn about a number of companies and the types of careers they offer. Job fairs also give you a chance to hand out your résumé and even interview and network on the spot.

you get started. For careers in the fields of architecture and engineering, a bachelor's degree is a necessity. For a career in construction, carpentry, and other more hands-on jobs in the building industry, experience is key. Do you need a college degree? Can you take a seminar or certificate program? Can you learn on the job? Researching the type of education needed for each area will help you understand the path you need to take to gain entry into the field.

- **Find an internship/volunteer opportunity:** Reading descriptions of jobs and profiles of college majors can provide only so much information as you explore career paths. An internship or volunteer opportunity will give you an up-close, hands-on look at what the career really entails. An internship or volunteer opportunity will give you the chance to handle assignments on your own and understand the day-to-day responsibilities required in the field. In theory, you might like the idea of building a boat. But when you understand the hard work associated with such a project—selecting wood or composite materials, shaping them to form the hull, crafting sails and rigging—hard reality may set in. Building a boat might not be what you had in mind after all. An internship or volunteer stint will help you discover all the ins and outs of a career field—both the good and the bad, the rewarding and the draining—and help you determine whether or not you're really cut out for the work.

## Scoring an Opportunity

Obtaining an internship or volunteer opportunity will take some work, but it will be time well spent. First, you'll want to identify existing opportunities. If you're a college student, your university's

# Top Internship Web Sites

- **Internships.com:** Offering internship guides, features, and articles, Internships.com connects students, employers, and educators to develop an abundance of national and international internship opportunities. The site provides a comprehensive and impressive number of internship opportunities often not found elsewhere.
- **Idealist.org:** More than 57,000 nonprofit and community organizations in over 180 countries list employment, internship, and volunteer opportunities on this site. There is a wide range of options for students who want to intern, volunteer, or work in positions that make a difference in the world.
- **Experience.com:** Offering internship listings and a wealth of resources and advice, Experience.com also features articles written by and for students who share their personal experiences.
- **GoAbroad.com:** GoAbroad.com is the umbrella organization that includes StudyAbroad.com, InternAbroad.com, and VolunteerAbroad.com. Thousands of international opportunities are listed for those who want an international internship, study, or volunteer experience.
- **Indeed.com:** This site includes job listings from major job boards, newspapers, associations, and company career pages. Indeed.com makes it possible to save searches and have internships and job postings sent to you via e-mail alerts.
- **Students.gov:** The official U.S. government Web site for college students offers a variety of internships and resources for those interested in interning or working for the U.S. government.

career services office can help. If you're in high school, you should start by making a list of local companies operating in your field of interest. For jobs in construction, you should seek out construction companies and independent contractors. Budding engineers should search for firms that specialize in the type of engineering in which they are interested. For example, civil engineers, who design roads, bridges, and other transportation infrastructure, can be found working for towns and cities or large firms.

Consult the Internet and search for companies in your geographical area. Also, look up company Web sites to learn more about their formal internship opportunities. A third option is to conduct an informational interview. An informational interview helps you learn about an occupation from someone who has firsthand knowledge. You interview a person who is doing a job that you find interesting. You can ask people questions about how they got

Your résumé is the single most important document you have to show off your knowledge, skills, and experience. A well-organized résumé that highlights your abilities will help you get noticed.

started, what their education is, what their day-to-day responsibilities are, and if any internship or volunteer opportunities exist in their company. You can also use the interview to identify to whom else you should be talking.

# Interviewing for an Internship

After you identify internship opportunities, the next step is to apply. You'll need to write a cover letter, create a résumé, and, if you get an interview, prepare for the type of questions they will ask.

Most interviewers begin by asking similar questions of each candidate, followed by questions specific to your experience. How you answer questions helps an interviewer understand who you are, what your personality is like, and how well you'll fit in with the company culture. Here are the most common interviewing questions and some possible ways to answer them:

**Q: Tell me a little about yourself.**

A: Don't give an interviewer your life story. What he or she really wants to know are facts about your education, your career aspirations, and current work or program of study.

**Q: Tell me what you know about this company.**

A: Do your homework before going on an interview. You should know about the company, its products or services, its leaders, and its customers. This will show that you are interested in what the company does and can talk about how your experience can help it in specific ways.

**Q: What relevant experience do you have?**

A: You should talk about all the experience you have gained either in classes or with projects and assignments you have completed that are relevant to the job you are interviewing for. You should also mention experience or leadership activities you've gained through sports or involvement in clubs.

Interviews can be intimidating, but they are an opportunity for you to show off what you know. A successful interview is one in which the applicant is well prepared to answer questions about his or her experience and skills.

## Q: What motivates you to do a good job?

A: Don't say money is a motivator—that's a turn-off to employers. Instead, focus on positive answers such as recognition for a job well done, becoming a leader in your field, making a meaningful contribution to the company and to the larger world, working within a team to achieve common goals, and becoming more knowledgeable about your job.

## Q: What's your greatest strength?

A: What this question is really asking is, "Why are you a great employee?" Here's where you can boast about what you have to offer. If you thrive under pressure, are a great motivator, take initiative, are a good problem solver, pay close attention to details, and are a hard worker, then say so and provide real-life examples and experiences to back up your claims. These are the kinds of qualities that employers want in a candidate.

## Q: What's your biggest weakness?

A: No one wants to admit to any weaknesses, so the strategy in answering this question is to turn a weakness into a positive. "Positive weaknesses" include being a perfectionist or being too committed to your job or caring too much.

## Q: Are you good at working in a team?

A: Always answer this question affirmatively. Projects are often team-based. You'll need to work with other team members to make the project cohesive and successful. This is also a great opportunity to talk about your leadership skills. Draw on your experiences at college with group projects or on your experience with summer jobs.

**Q: Do you have any questions to ask me?**

A: You want to be sure to ask questions of the interviewer. Asking what your duties will be, who you will work under, what the expectations are, or how you will be evaluated will give you some idea about whether you think the job is a good fit for you. Interviews are not just about an employer choosing the right candidate for a job; they are also an important opportunity to determine if the company is a place where you can thrive and excel. It's not only about if you are right for the company, but also if the company is right for you.

# Chapter Three

# OPPORTUNITIES IN CARPENTRY

For people who love to build things, carpentry might be the most diverse of the building career fields. Carpenters work in many different areas within the larger construction industry. In fact, within the greater construction field, carpenters make up the highest number of self-employed people. Well-rounded carpenters with diverse skills are in demand.

Job duties can include cutting, sizing, and constructing wood and other materials for homes, businesses, factories that make products (like guitars, furniture, or cabinets), and cities and towns (e.g., public piers, bridges). Carpenters who are employed by a building contractor for housing or commercial projects may be asked to perform a number of different jobs like wall-framing, installing windows and doors, constructing stairs, laying floors, making cabinets, or fitting moldings.

A carpenter's work varies depending on where he or she works and what is expected of him or her. A carpenter working at a home might help build framing for a wall, install windows, or build stairs or a deck. One who works for a musical instrument company might select wood and materials based on how they influence sound when a completed

Carpentry is a diverse and exciting career field in which skilled workers create everything from cabinets and shelving to musical instruments and wooden boats.

instrument is played. Or the carpenter may craft an instrument by hand, tailored to a musician's unique specifications.

## Internship Overview

Carpentry is a career field in which individuals learn their skills on the job. Though there are programs offered at vo-tech (vocational/technical) schools, most carpenters learn through apprentice programs, which are similar to internships. Most of an apprentice's training is done while working for an employer who helps the apprentice learn the trade or profession. Apprenticeships offer the most comprehensive training. Local unions like the United Brotherhood of Carpenters and Joiners of America, the National Association of Home Builders, and the Associated General Contractors of America all sponsor apprenticeships or training programs.

Furniture makers are craftsmen who build a variety of furniture for display, function, and comfort. They are highly skilled workers who design, construct, and repair furniture or wooden products for homes and businesses. Furniture makers create furniture like tables, chairs, cupboards, chests, drawers, and office furniture. Furniture makers can also repair damaged furniture and restore antiques. Internships in furniture-making teach students the basics of woodworking, preparation and finishing, joinery, use of proper fastenings and findings, use of carpentry tools, cabinetmaking, refinishing, and refacing. They also learn the history of furniture design, contemporary furniture design, and drawing digital designs.

Boat building consists of skills and techniques specific to crafting wooden boats. Building traditional wooden boats requires specialized skills such as design and lofting, timber harvesting, bronze casting, and all of the techniques of tool use and methods of plank-on-frame boat building.

## Opportunities Available

Carpentry internships are often called apprenticeships. Preparation for a carpentry apprenticeship should begin in high school with classes in wood shop or in a vo-tech course. Professional carpenters expect an apprentice to have some knowledge of the field before he or she begins.

- **Furniture making:** Some internships are offered by furniture companies, such as Stickley, Audi & Co., and Ethan Allen. Others can be found at art centers or design institutes specializing in fine art furniture or antique restoration.

Alexandria Seaport Foundation executive director Joe Youcha explains boat-building details to apprentice Oscar Melgar. The foundation teaches high school and GED students math skills through boat building in an effort to get more trained workers into construction jobs.

- **Boat building:** Some internships in the boat building industry focus on preservation efforts. These interns learn to preserve ship objects made of different materials, including wood, iron, metal, leather, glass, and ceramic. Some interns may even have the opportunity to assist in restoring an antique boat while learning traditional, premechanized building and repair techniques.
- **Wooden instrument making:** Making a wooden musical instrument can be an intensive process. A luthier, or a person who specializes in making violins or guitars, needs to be educated in the different types of woods used to make these instruments; how to shape wood to achieve specific acoustical qualities; the installation of strings, frets, and fingerboards; and finishing the wood to achieve a particular look.

## Typical Tasks and Skills

Becoming a carpenter involves a number of qualifications. A high school diploma isn't required but is a valuable asset, especially since carpenters use math skills on a daily basis. Good carpenters also have strong hand-eye coordination and are physically strong.

Carpentry apprenticeships teach students the skills and knowledge necessary to become a professional carpenter. An apprentice begins by studying basic math, measurements, and other fundamental technical and industrial skills. The apprentice program goes on to offer instruction in everything from measuring and cutting wood and selecting the right types of wood for a project, to more advanced techniques in joinery (or joining together pieces of wood), types

Carpentry is a hands-on field, where interns learn by completing actual projects. They will also work with specialized equipment as they learn to make accurate measurements, cut materials, join them together, and sand and finish.

of joinery materials (fasteners, bindings, or adhesives), sanding and finishing, and the use of power tools. An apprentice program teaches about building materials such as the various types of woods and which ones are best for crafting furniture versus those used for framing a house.

Apprentice carpenters will learn how to read drawn and illustrated plans and how to edit plans. They will also learn the basic rules of the National Building Code. This is a set of rules that specify the minimum acceptable level of safety for constructed buildings and structures.

# What You'll Have Learned

After completing an apprenticeship in carpentry, students will have learned to:

- Use basic mathematics (algebra, geometry, and trigonometry)
- Apply concepts for the reading of drawings used in the fields of construction and maintenance
- Classify different types of wood and learn which ones are best used for particular purposes
- Describe the tasks involved in roof and stair building
- Select wood for millwork, and specify moldings and types of joints for cabinet work

# Chapter Four

# OPPORTUNITIES IN CONSTRUCTION

**H**ouses, apartments, factories, offices, schools, roads, and bridges are just some of the projects a construction company handles. There are two distinct segments to the construction industry: commercial and residential.

Commercial construction consists of building any structure that contains a business. High-rise buildings; recreational, institutional, and government facilities; public works projects; retail/commercial outlets; and office facilities are just a few examples of commercial projects a construction company might build. Residential construction is any building in which the structure being built will be used as a home after its completion.

## Internship Overview

Internships in the construction industry enable a student to actively participate in the entire building process. Construction interns will gain a variety of knowledge about the construction industry as a whole, as well as detailed knowledge of how a building is built.

High above Manhattan, atop a skyscraper still being built, this construction worker builds a wooden frame that will support a deck upon which a concrete floor will be poured.

There are three areas that a construction intern should understand. They are field engineering, project management, and acquisition.

Field engineering provides an understanding of how a building is built, from the ground up. Students learn about job sites, what it takes to get construction started, and the coordination efforts needed. Project management gives you an understanding of a project's day-to-day elements, as well as the long-term business and financial management components of a construction company. This information is key if you want to eventually own your own construction company. Acquisition consists of the necessary tasks that need to be completed prior to the start of a construction project. These can range from preconstruction, estimating a job, and purchasing materials.

## Opportunities Available

Construction interns can gain experience in a number of different places, from large construction firms to smaller specialty contractors. A larger firm will give you a broad overview of the entire construction process, with an emphasis on project management. A smaller firm or specialty contractor might give you a more hands-on experience where you'll learn to work with particular materials and personally build portions of a job.

Many of these opportunities are only available to students who have completed one-to-two

These construction workers are building a handicap ramp for a sidewalk. Construction workers can find employment at major construction firms, small companies specializing in home improvement projects, or with the public works departments of cities and towns.

# Working for a City Department of Public Works

A city's department of public works is made up of a staff of employees who are responsible for the construction, renovation, and operation of the city's facilities and infrastructure. These departments employ carpenters, contractors, and engineers who oversee the building and maintaining of the city's streets, installation of its sewers, construction of storm drains, building and inspecting of bridges, planning and development of public park areas, and construction of public buildings and service facilities. They often contract with carpenters and other professionals who love to build things to work on a variety of these projects.

years of college coursework. The nature of these opportunities often requires a minimum level of knowledge and experience that high school students do not yet possess. However, some companies do feature internship or shadow programs for high school students to learn more about the construction industry.

## Typical Tasks and Skills

A construction internship will give you hands-on experience in the field. You might perform some of the following tasks during a construction internship:

- Assist project managers
- Coordinate subcontractors and their activities
- Manage scheduling, process work orders, and prepare requests for subcontractors' payments

# Habitat for Humanity

Habitat for Humanity is a nonprofit organization that assists in building affordable housing for people in need in communities in the United States and around the world. Homes are built by volunteers from the community who contribute their skills to the project. As a volunteer for Habitat for Humanity, you can learn about framing, masonry, siding, roofing, painting, finish work, and other construction skills needed to build a house. The organization welcomes everyone, regardless of experience level. Habitat volunteers work on a team composed of both experienced builders and people who just want to contribute to their community but have few skills. Less knowledgeable volunteers learn on the job from team members with extensive experience in construction and home building.

For a broader building experience, you can apply to become a Habitat for Humanity Global Village volunteer. These volunteers work on housing construction and renovations and assist in disaster-relief efforts in diverse international locations such as Kenya, Cambodia, Fiji, New Zealand, Brazil, El Salvador, Romania, and Thailand. You and other Global Village volunteers will work alongside members of the community in which you are placed to build homes and assist in fund-raising.

- Obtain the appropriate and necessary building permits and licenses
- Check zoning and building code requirements
- Design temporary facilities
- Assist in conducting safety inspections
- Document material deliveries
- Conduct site tours and job site walks
- Update database files with work hours and materials used
- Prepare reports and presentations

## What You'll Have Learned

You can expect to learn a number of important skills as an intern in construction. These include:

Habitat for Humanity is an international organization that has built more than five hundred thousand homes around the world for families in need.

- Understanding of building regulations
- Measuring, marking, and organizing materials based on a blueprint
- Sizing, forming, and cutting materials from wood and fiberglass to drywall, tile, or other types of flooring
- Using a variety of tools, such as chisels, hammers, screwdrivers, sanders, saws, and drills
- Connecting the pieces of a project using adhesives or nails, depending on the project or materials used

# Chapter Five

# OPPORTUNITIES IN ARCHITECTURE AND DESIGN

**A**rchitect David Childs was faced with an unprecedented challenge. Two years after the terrorist attacks that destroyed the World Trade Center in New York City on September 11, 2001, he was asked to design a new skyscraper on the Ground Zero site. The building needed to honor those lost in the attack, be a symbol of hope, exhibit the latest in security and safety measures, and be cost effective—all while looking elegant and fitting in with the iconic New York City skyline.

Childs' task is a common one for architects. While he faced the added pressure of his building serving as a symbol of New York's resilience, his project also showed the many different elements every architect has to manage when designing a building.

## Internship Overview

Architects are primarily responsible for designing, planning, and overseeing the physical construction of any number of structures. Architects design skyscrapers stadiums and sports facilities, schools, corporate business

Architectural drawings, blueprints, and models are used to show what a building will look like once it's erected.

parks, amusement parks, malls, recreation areas, movie theaters, gas stations, restaurants, houses, and apartment buildings. Architects combine design and engineering skills with a knowledge of physics and construction. It is a creative field where a structure can look modern and cutting-edge or be a throwback to a bygone era. While many office towers and high-rise hotels use steel and glass to look sleek and contemporary, sports parks are often designed to resemble classic stadiums from the first half of the twentieth century, while also adding innovative touches.

Architecture interns can gain experience by working in small or large architecture firms or with engineers, contractors, and other building professionals in firms across the country and around the world. An architecture intern supports architects or design professionals, helps revise design plans, builds scale models, and assists with presentations to clients.

## Opportunities Available

There are many opportunities for internships in architecture. Architectural firms, such as HOK (which designed the Indianapolis International Airport), Perkins + Will (the San Francisco Conservatory of Music), and Gwathmey Siegel Kaufman & Associates Architects (the Euro Disney

Architecture interns learn how to design, plan, and oversee the construction of a building, park, home, or any number of structures. Interns will also witness decision-making processes, learn how projects are managed, and see how other professionals (such as construction workers, woodworkers, interior designers, electricians, and plumbers) work together to complete a project.

Convention/Exhibition Center and Hotel in France) often offer internships to college students majoring in architecture. Many of these opportunities are available to students who have completed one to two years of college coursework or who have begun advanced studies at the graduate level.

An emerging specialty within architecture is green architecture, often referred to as sustainable design or eco-friendly building. Green architecture is the construction of buildings using environmentally conscious design techniques and materials that lessen a building's carbon footprint, minimizing its negative environmental impact. This area of architecture is fairly new, but internships do exist for students interested in understanding how the latest techniques and materials have positively impacted both architecture and the environment.

Internships at many of these architecture and design firms are gateways to future employment. The internship gives the company a chance to see what type of employee you will be and how

## Architecture Is About Teamwork

The September 11, 2001, attacks changed how office towers are designed, built, and protected. When David Childs was designing the new skyscraper at Ground Zero, he collaborated with police, fire officials, and security experts, as well as other architects, developers, and engineers. This collaboration and consultation resulted in the creation of a design that incorporated the newest safety standards and the most high-tech security features, as well as the most beautiful aesthetics. The new building can't look like a fortress, but, as Childs says, "It has to be a great place to work, and it has to be safe."

well you can contribute to the company, if you'll be a team player, and how well you can apply your skills to the real-world projects to which you will be assigned.

## Typical Tasks and Skills

Internships for students interested in architecture will often help refine the skills needed for success in the field. For example, architecture interns will gain hands-on experience planning and designing buildings, landscapes, and other common projects. They will draw their ideas out in diagrams that show the relationship between a building and the spaces around it, such as

## The New Yankee Stadium

When architects sat down to create a design for the new home of the New York Yankees baseball team, they looked to the past as much as the future. The original Yankee Stadium, constructed in 1923, featured a gently curving frieze that crowned the upper deck, which had become a distinguishing characteristic of the stadium and the team. The frieze has even become a trademark graphic element used on T-shirts, baseball caps, and other Yankee memorabilia. The new stadium sought to reference and honor this iconic architectural element of the old stadium. Made from copper, which turns green over time from exposure to air, the new frieze is not merely a decorative element. It also has an architectural, load-bearing function. It is part of the support system for the cantilevers, or beams, beneath the upper deck, and for the lighting above it.

freeways, roads, mountains, and parking lots. They will present their work to clients, and they will manage a project to ensure that what is being built meets their design specifications.

## What You'll Have Learned

Students should come away from an architecture internship with much of the knowledge and skills they will need to be successful in the field. An architectural internship will give you the opportunity to assist a senior-level architect, who will give you an understanding of architectural planning, development, design, and production. You'll also witness decision-making processes, learn how complex projects are managed, understand the role that construction and construction management plays, and participate in the work culture of architects and engineers.

Architecture interns will work with senior-level architects as they gain an understanding of planning, development, and design.

After you complete an architectural internship, you will have gained skills in:

- Understanding architecture and its role in society
- Design and drawing
- Communication and documentation
- Problem solving
- Teamwork

# Chapter Six

# OPPORTUNITIES IN ENGINEERING

**A** team of engineers, technicians, and scientists monitor the Hubble Space Telescope 24-hours-a-day, 365 days a year. So when a problem occurs, like when the telescope stops sending information to Earth, the team responds quickly by troubleshooting what might be wrong. For the engineers who work on the telescope, testing and maintaining the spacecraft's overall performance is important. Whenever the telescope stops transmitting data, engineers reboot the various systems that operate the telescope to get it back online.

Engineering is an exciting job that combines science, technology, and imagination to build innovative pieces of machinery, from the Hubble Telescope to roller coasters, automobiles, planes, and satellites.

While engineers do not physically build the things they design, they do use their skills to create things such as roller coasters and cars. They do this through the art and science of design. An engineer might design a new car and then hand the plans off to a manufacturer, who actually

Samantha McCue, a student at the University of Illinois, completed an internship at NASA. Sixty percent of NASA's job opportunities are in the engineering and engineering-related fields.

creates and assembles the components of the vehicle. An engineer might oversee the entire process to be sure the car is built to the engineer's specifications.

## Internship Overview

Many college engineering programs require students to complete an internship or a cooperative education assignment as part of their education. These opportunities give students a chance to apply what they have learned in the classroom while helping to solve real-world engineering problems facing actual companies.

Engineering internships are a great way to gain hands-on experience working on a real-world project. Interns work directly with engineers and other engineering professionals in a team environment. This approach shows students how an engineer works with professionals from other fields on a project and what an engineer's role is in relation to other members of the project's team. Interns also get a chance to develop their skills under the supervision and guidance of experienced engineers, who can give advice on career goals and educational aspirations while sharing their own job experiences.

# Types of Engineers

**Civil Engineers:** Civil engineers study all of the aspects involved in creating new buildings, bridges, highways, roadways, and other types of infrastructure. This includes the planning, designing, constructing, operating, and maintenance of the infrastructure. They also improve existing infrastructure that has been neglected. Civil engineers usually practice in a particular specialty, such as construction engineering, geotechnical engineering, structural engineering, land development, transportation engineering, hydraulic engineering, and environmental engineering. Some civil engineers, particularly those working for government agencies, may practice multiple specializations, particularly when involved in critical infrastructure development or maintenance.

**Computer Engineers:** These engineers design programs and new technologies to make computers faster, more efficient, more effective, and more powerful. They integrate several fields of electrical engineering and computer science required to develop computer systems. Computer engineers usually have training in electrical engineering, software design, and hardware-software integration, instead of only software engineering or electronic engineering. Computer engineers are involved in many hardware and software aspects of computing, from the design of individual microprocessors, personal computers, and supercomputers, to circuit design. This field of engineering not only focuses on how computer systems themselves work, but also how they integrate into the larger picture.

**Electrical Engineers:** Electrical engineers study and apply their knowledge of electricity, electronics, and electromagnetism to create a wide variety of digital signal processors, microcontrollers, computing devices, transformers, electric generators, electric motors, high voltage engineering, and power electronics. Electrical engineers can specialize in a range of subtopics including power, electronics, control systems, signal processing and telecommunications. They can be involved with everything from large-scale electrical systems such as power transmission and motor control, to the study of small-scale electronic systems including computers and integrated circuits.

**Industrial Engineers:** Industrial engineers are primarily concerned with the optimization of complex processes or systems. They focus on the development, improvement, implementation, and evaluation of integrated systems of people, money, knowledge, information, equipment, energy, materials, and analysis and synthesis. They integrate the mathematical, physical, and social sciences with the principles and methods of engineering design to specify, predict, and evaluate the results to be obtained from such systems or processes. Industrial engineering is also known as operations management, management science, operations research, systems engineering, manufacturing engineering, or ergonomics and human factors engineering/safety engineering.

**Mechanical Engineers:** Mechanical engineers apply the principles of physics and materials science for analysis, design, manufacturing, and maintenance of mechanical systems. Mechanical engineering is the branch of engineering that involves the production and usage of heat and mechanical power for the design, production, and operation of machines and tools. The engineering field requires an understanding of core concepts including mechanics, kinematics, thermodynamics, materials science, and structural analysis. Mechanical engineers use these core principles along with tools like computer-aided engineering and product lifecycle management to design and analyze manufacturing plants, industrial equipment and machinery, heating and cooling systems, transport systems, aircraft, watercraft, automobiles, roller coasters, satellites, missiles, robotics, and medical devices such as prosthetic and artificial body parts, pacemakers, and incubators. Mechanical engineers must stay on top of developments in such fields as composites, mechatronics, and nanotechnology. Mechanical engineering often overlaps with aerospace engineering, building services engineering, civil engineering, electrical engineering, petroleum engineering, and chemical engineering.

# Opportunities Available

Opportunities for engineering internships or cooperative education assignments are plentiful. Many corporations, including car companies like Toyota, Honda, General Motors, and Tesla Motors, as well as aerospace and spacecraft agencies like NASA, Pratt & Whitney, and Lockheed Martin, offer internships to students.

Many of these opportunities are only available to students who have completed one to two years of college coursework. The nature and demands of the work often require a minimum level of knowledge and experience that high school students do not yet possess. However, some companies do have internship or shadow programs for high school students to learn more about the career field before they select engineering as a college major.

Internships at many of these companies are gateways to future employment. The internship gives the company a chance to see what type of employee you will be and how well you can contribute to the company, if you'll be a team player, and how well you can apply your skills to the projects you'll be assigned to.

# Typical Tasks and Skills

In an engineering internship/co-op, you will apply the skills you learned in the classroom to real-world projects, thereby experiencing how industrial problem solving works. You will also develop leadership abilities, learn how to improve your productivity, and understand the role of engineering within the company and in society at large. As part of your internship experience, you'll receive hands-on learning and participate in actual projects where your contributions will be vital to success. You will also be exposed to the professional manufacturing environment.

An internship will help you develop the skills you need to be a successful engineer, such as:

Ryan Joyce, a student at the Cleveland Institute of Art, poses with one of his designs at his home. Joyce's work during his General Motors internship will be featured in a Discovery Channel documentary titled *Futurecar*.

- **Strong Analytical Aptitude:** An engineer has excellent analytical skills and is continually examining things and thinking of ways to make things work better.
- **Attention to Detail:** Engineers pay careful and meticulous attention to detail. The slightest error can cause an entire structure to fail. Every detail must be reviewed thoroughly during the course of completing a project.
- **Communication Skills:** Engineers can speak plainly and accessibly about complex technical projects when they need to discuss something with a layperson, such as a client or non-engineering professionals working on the same project.
- **Lifelong Learning:** Because engineering is a technology-based field, advances in materials, techniques, tools, and capabilities occur regularly. Engineers must be lifelong learners to stay on top of the latest developments and advances.
- **Creativity:** Creative and innovative ways of doing things are often the result of engineers who think outside the box.
- **Logic:** Making sense of complex systems and understanding how things work is the main objective of an engineer.
- **Mathematical Inclination:** Engineering involves complex calculations that need to be precise.
- **Problem Solving:** Engineers address problems, so they must be able to ascertain where a problem is coming from, what might be causing it, and how to create an efficient and smart solution.

- **Teamwork:** Engineers understand that they are part of a larger team working together on a project. Working well with different types of people and valuing their contributions are key.
- **Technical Knowledge:** Technical knowledge is a must for an engineer. He or she must know his or her area of engineering (such as mechanical or civil), but also have a basic understanding of many other areas, such as computer engineering, electrical engineering, manufacturing engineering, and industrial engineering, since these areas all come together in a single project and impact one another.

## What You'll Have Learned

In an engineering internship or co-op, you'll have learned how the various types of engineers work together to design, build, complete, and operate a project, and how engineers work with other members of the team—such as construction companies, manufacturers, architects, and clients.

Chris Allen, who attended the Oregon Institute of Technology, completed an internship at Boeing. That experience led to a full-time job with the company in manufacturing engineering.

# Chapter Seven

# MAKING THE MOST OF YOUR INTERNSHIP OR VOLUNTEER OPPORTUNITY

O nce you have landed your internship or volunteer opportunity, you'll want to be sure you use the experience to your advantage. Here are the best ways to make sure you are getting the most out of your work experience:

- **Choose the right internship/volunteer experience:** Think about what you want out of your experience and tell your supervisor what you are hoping to gain during your time with the company. Do you want to learn a specific set of skills? Do you want to contribute to a project so that you can list it on your résumé? Be clear about what you hope to gain from your experience.
- **Define the job's responsibilities:** You want a work experience that gives you hands-on experience in your career field, not one where you'll just get coffee or run errands for the key members of your team. Ask upfront what you'll

Internships and volunteering are a great opportunity to learn new building and construction skills, enhance your experience, and even try something new.

do day-to-day, if you can participate directly on projects, and how your work will be evaluated. Ask a lot of questions.

- **Make a good impression:** It sounds simple, but be responsible, show up on time, work hard, act professionally, take initiative, be self-sufficient, and do your best on each assignment. Doing these things will go a long way toward proving that you are professional, can handle the work, are a team player, can work independently, and are worth hiring later on.

- **Use your experience to launch your career:** If you'd like to work at the company after graduation or after you have completed your internship/volunteership, stay in touch with your supervisor and others you've met at the company. Keep them informed of your progress in school, other internships you have completed, school projects they might find interesting, or other work experiences you have had that might make you more valuable to them. Also, be sure to keep copies of your evaluations, reports you've written, presentations you've worked on, projects you've completed, or other assignments you took part in during your internship or volunteer opportunity.

# Tips for Making the Most of Your Work Experience

1. **Be early.** Always arrive early for work each day and to meetings. It shows you value your internship or volunteer opportunity and you are respectful of others' time. What will you do each day? To whom will you report? How will you be evaluated? Find out these key pieces of information so that you can make the most of your time on the job.

2. **Visit the break room and water cooler.** A lot of project decisions are made and office networking is conducted in break rooms or around the water cooler. Don't be shy about saying hello to your boss, the head of human resources, or anyone else who might be able to help you land a job later on.

3. **Don't get discouraged.** If you are performing tasks you don't like or don't feel like you are contributing any truly important work to the company, don't worry. Work ebbs and flows between busy and slow periods. Also, once a supervisor feels he or she can trust you with smaller tasks, he or she will begin offering you more challenging and rewarding assignments.

4. **Don't stand out.** Some interns think this is their time to shine by showing off all their skills. Resist the urge to show off. Many companies want to see if you fit smoothly and comfortably within the corporate culture. Your work will speak for itself without you loudly calling attention to it.

5. **Take notes.** Record everything you work on during your internship or volunteer opportunity. You may only be there a short time, and, while you think you'll remember everything, you most likely won't. Keeping notes will also help you add to your résumé later by including details about your involvement in various projects.

6. **Get involved.** Join the company softball league, go on that coffee run (even if you don't drink coffee), or tag along on a client meeting. Your supervisor wants to know how you act outside of work, if you are personable and gregarious, and if you can handle anything that's thrown at you.

## Learning from the Experience

After you have completed your internship, hopefully you will have a better understanding of your career field. But you should also have an idea of the skills you must acquire to be successful, any educational requirements you need to have, or additional training you might want to obtain.

For students returning to college after an internship, this is a great opportunity to examine your program of study to determine if there are any holes you need to fill. For example, were there professionals that you worked with who had skills you want or need to learn? Seek out additional courses or see if there are seminars you can take to help you gain those skills. Did you find during your internship that you're in the wrong major or should pick up a second major? Consult an academic adviser to see if you should switch programs or make other alterations and adjustments to your program of study.

For those who used the internship to learn the skills needed for a chosen career, now's the time to explore whether there are additional skills you need to have that you don't currently

Job counselors and career specialists can assist students in defining their interests and matching them with a field that fits their career goals.

possess. A college course, seminar, or other opportunity to learn them is a great way to meet more people in your field and make connections with other professionals.

## After the Internship/Volunteer Experience Is Over

After you have completed your internship or volunteer opportunity, hopefully you will have gained a better understanding of your career field and the skills and education needed to be a success.

For internships or volunteer opportunities that are completed during college, or as part of a program's educational requirements, students will want to return to school to complete their education. Make sure you keep in touch with your supervisors or other professionals you worked with during your time at the company or organization. They will be the key to helping you get a job later. They may serve as references, sources of information, or someone who can point you toward job openings in the field. Social networking (Facebook, Twitter, etc.) and e-mail are easy ways to stay in touch with your professional contacts.

For those in career fields such as construction, where an internship or apprenticeship teaches you hands-on skills that can only be acquired through actual work experience, the next step is to begin looking for your first full-time job. Begin by updating your résumé to reflect your internship/apprenticeship, listing and briefly describing the projects you worked on and your role in their success. Next, you'll want to leverage the contacts you've made and relationships you've formed during your internships. Ask the people you worked for, such as your supervisor, if they will serve as references for you. They will be able to speak directly to your skills when a future employer asks for their thoughts on your work ethic and contributions.

After an internship experience is over, students are well positioned to apply for additional internships, apprenticeships, or employment in their field of interest. Many students enter the workforce or continue their education.

# I Hated My Internship/Volunteer Opportunity: Now What?

The value of an internship or volunteer opportunity is learning about the career field you have an interest in. But what if the work experience and the field turn out to be not what you expected or hoped? Sometimes when we see exactly what is involved in the day-to-day aspects of a job, we find we dislike it or that it doesn't match our interests and skills. There is value in learning this early on. By discovering this sooner rather than later, you have time to change your mind, explore other options, or switch your major to something else that interests you more and better suits you.

The next step is to restart your career search by talking to a career counselor or your academic adviser (if you attend college) and researching jobs on the Internet. Now that you know what you dislike doing, you can use that information to narrow down your search. You can now focus more closely on the things you know you do like to do and can do well.

# Glossary

**acoustic**  Music produced solely or primarily with instruments that make sound without the aid of electricity.

**apprenticeship**  A system of teaching a new generation of practitioners a competency-based set of skills within a structured setting.

**buttress**  An architectural structure built against or projecting from a wall that serves to support or reinforce the wall.

**composite**  Material made from two or more significantly different materials that are combined to create one material.

**credit**  A unit of classroom time or coursework that is a requirement leading to an academic degree.

**density** The mass of a substance per unit of volume; the distribution of a quantity per unit of space.

**drywall**  Material used to make a building's interior walls and ceilings.

**framing**  A building technique based around structural members, usually called studs, which provide a stable frame to which interior and exterior wall coverings are attached.

**hexagon**  In geometry, a shape that features six edges and six vertices.

**hull**  The watertight body of a ship or boat.

**infrastructure**  The basic physical and organizational structures needed for the operation of a society or enterprise, such as roads, bridges, tunnels, sewers, electrical grids, and subways.

**joinery**  Woodworking joints or other types of mechanical joints.

**luthier**  Someone who makes or repairs stringed instruments.

**restoration**  To bring something back, through repair and
refurbishment, to its original condition.

**rigging**  A system of ropes, chains, and tackle used to support
and control the masts, sails, and yards of a sailing vessel.

**piledriver**  A mechanical device used to drive piles into soil to
provide foundation support for buildings, piers, or other
structures.

# For More Information

American Institute of Architects
1735 New York Avenue NW
Washington, DC 20006-5292
(202) 626-7300
Web site: http://www.aia.org
This has been the leading professional membership association
for licensed architects, emerging professionals, and allied
partners since 1857.

Architectural Engineering Institute
1801 Alexander Bell Drive
Reston, VA 20191-4400
(800) 548-2723
Web site: http://www.asce.org/aei
The Architectural Engineering Institute is a professional
organization for architectural engineers.

Connect Canada
401 Sunset Avenue
Windsor, ON N9B 3P4
Canada
(519) 253-3000, ext. 4130
Web site: https://www.connectcanadainternships.ca
Connect Canada is a national internship program that links
Canadian companies with graduate students and
postdoctoral fellows for research placements. Regardless of
the sector or the location, Connect Canada interns will
apply the latest in research methods and know-how to the
most pressing industrial research and development (R&D)
issues. Companies receive a cost-effective way to conduct

R&D, while interns gain relevant industry experience that enhances their graduate studies.

Construction Management Association of America
7926 Jones Branch Drive, Suite 800
McLean, VA 22102-3303
Web site: http://cmaanet.org
This is North America's only organization dedicated exclusively to the interests of professional construction and program management.

Corporation for National and Community Service
1201 New York Avenue NW
Washington, DC 20525
(202) 606-5000
Web site: http://www.serve.gov
The Corporation for National and Community Service is a federal agency that engages more than five million Americans in service through Senior Corps, AmeriCorps, and Learn and Serve America. It also leads President Barack Obama's national call-to-service initiative, United We Serve.

International Volunteer Programs Association
P.O. Box 287049
New York, NY 10128
Web site: http://www.volunteerinternational.org
The International Volunteer Programs Association is an association of nongovernmental organizations involved in international volunteer work and internship exchanges. It's an association of volunteer-sending organizations but does not organize or run its own volunteer programs.

Internship Institute
2865 South Eagle Road

Newtown, PA 18940

(215) 870-9700

Web site: http://www.internshipinstitute.org

The Internship Institute is a nonprofit, nonpartisan organization whose mission is to assure the quality, integrity, and success of internships in order for individuals, organizations, and economies to prosper. It is the only nonprofit solely dedicated to the advancement of best practices globally through education, collaboration, and advocacy.

National Trust for Historic Preservation

1785 Massachusetts Avenue NW

Washington, DC 20036-2117

(800) 944-6847

Web site: http://www.preservationnation.org

This organization provides leadership, education, advocacy, and resources to save America's diverse historic places and revitalize its communities.

SkillsUSA

1601 Jefferson Street

Alexandria, MN 56308

(320) 762-0221

Web site: http://www.alextech.edu/en/students/programs/ Manufacturing/Carpentry/SkillsUSA.aspx

SkillsUSA is a student organization known for supplying the workforce with highly qualified carpenters.

United Brotherhood of Carpenters

395 Hudson Street

New York, NY 10014

(212) 366-7500

Web site: https://www.carpenters.org

The United Brotherhood of Carpenters is one of North America's largest building trades unions, with nearly a half million members in the construction and wood-products industries.

Volunteer Canada
353 rue Dalhousie Street, 3rd Floor
Ottawa, ON K1N 7G1
Canada
(613) 231-4371
Volunteer Canada is committed to developing involved Canadians, resilient communities, and a vibrant Canada. It seeks to provide leadership in strengthening citizen engagement and serving as a catalyst for voluntary action.

VolunteerMatch
717 California Street, 2nd Floor
San Francisco, CA 94108
(415) 241-6868
VolunteerMatch strengthens communities by making it easier for good people and good causes to connect. The organization offers a variety of online services to support a community of nonprofit, volunteer, and business leaders committed to civic engagement. Its popular service welcomes millions of visitors each year and has become the preferred Internet recruiting tool for more than eighty-one thousand nonprofit organizations.

# Web Sites

Due to the changing nature of Internet links, Rosen Publishing has developed an online list of Web sites related to the subject of this book. This site is updated regularly. Please use this link to access the list:

http://www.rosenlinks.com/FID/Build

# For Further Reading

Berger, Lauren. *All Work, No Pay: Finding an Internship, Building Your Résumé, Making Connections, and Gaining Job Experience.* Berkeley, CA: Ten Speed Press, 2012.

Berger, Sandra. *Ultimate Guide to Summer Opportunities for Teens: 200 Programs That Prepare You for College Success.* Waco, TX: Prufrock Press, 2007.

Byers, Ann. *Jobs as Green Builders and Planners.* New York, NY: Rosen Publishing, 2010.

Ferguson Publishing. *Careers in Focus: Construction.* New York, NY: Ferguson Publishing, 2010.

Fraser, Aime Ontario, Joe Hurst-Wajszczuk, and Matthew Teague. *Woodworking 101: Skill-Building Projects That Teach the Basics.* Newtown, CT: Taunton Press, 2012.

Gay, Kathlyn. *Volunteering: The Ultimate Teen Guide.* New York, NY: Scarecrow Press, 2007.

Hindman, Susan. *Carpenter* (Cool Careers). New York, NY: Gareth Stevens Publishing, 2010.

Hutson, Matt. *Totally Amazing Careers in Engineering.* New York, NY: Sally Ride Science, 2007.

Landis, Raymond. *Studying Engineering: A Roadmap to a Rewarding Career.* Geelong, Australia: Discovery Press, 2007.

Lyden, Mark. *College Students: Do This! Get Hired!: From Freshman to Ph.D.: The Secrets, Tips, Techniques, and Tricks You Need to Get the Full-Time Job, Co-op, or Summer Internship Position You Want.* Charleston, SC: BookSurge Publishing, 2009.

McDavid, Richard A. *Career Opportunities in Engineering.* New York, NY: Checkmark Books, 2007.

Mondschein, Kenneth C. *Construction and Trades* (Great Careers with a High School Diploma). New York, NY: Ferguson Publishing, 2008.

Peterson's. *Teen's Guide to College & Career Planning: Your High School Roadmap for College & Career Success* (Teen's Guide to College and Career Planning). Lawrenceville, NJ: Peterson's, 2008.

Roza, Greg. *A Career as a Carpenter.* New York, NY: Rosen Classroom, 2010.

Senker, Cath. *Construction Careers.* Mankato, MN: Amicus, 2011.

Sheldon, Roger. *Opportunities in Carpentry Careers.* New York, NY: McGraw-Hill, 2007.

Sumichrast, Michael. *Opportunities in Building Construction Careers.* New York, NY: McGraw-Hill, 2007.

Waldrep, Lee W. *Becoming an Architect: A Guide to Careers in Design.* Hoboken, NJ: Wiley Publishing, 2009.

Woodard, Eric. *Your Last Day of School: 56 Ways You Can Be a Great Intern and Turn Your Internship into a Job.* Seattle, WA: CreateSpace, 2011.

# Bibliography

Agovino, Theresa. "Redefining New York City's Skyscrapers."
    *Crain's New York Business*, May 8, 2011. Retrieved
    February 2012 (http://www.crainsnewyork.com/
    article/20110508/9_11/305089979#).

Berger, Lauren. *All Work, No Pay: Finding an Internship, Building
    Your Résumé, Making Connections, and Gaining Job
    Experience.* Berkeley, CA: Ten Speed Press, 2012.

Bolles, Richard N., Carol Christen, and Jean M. Blomquist. *What
    Color Is Your Parachute for Teens: Discovering Yourself,
    Defining Your Future.* Berkeley, CA: Ten Speed Press, 2006.

Career Overview. "Carpenter and Carpentry Careers, Jobs and
    Training Information." Retrieved March 2012 (http://www
    .careeroverview.com/carpenter-careers.html).

Chambers, Erin. "Tips to Make the Most of Summer Internships."
    *Wall Street Journal,* July 1, 2008. Retrieved March 2012
    (http://online.wsj.com/article/SB121450293562107717.html).

Dream Careers. "Architecture Internships." Retrieved
    March 2012 (http://www.summerinternships.com/
    architecture-internships).

EngineeringSchools.com. "10 Qualities of a Great Engineer."
    Retrieved March 2012 (http://engineeringschools.com/
    resources/top-10-qualities-of-a-great-engineer).

Griepentrog, Troy. "A Fun Way to Learn DIY Skills and Benefit
    Your Community." *Mother Earth News*, March 6, 2008.
    Retrieved March 2012 (http://www.motherearthnews.com/
    Do-It-Yourself/Volunteer-For-Habitat-for-Humanity.aspx).

Liang, Jengyee. *Hello Real World!: A Student's Approach to
    Great Internships, Co-ops, and Entry-Level Positions.*
    Charleston, SC: BookSurge Publishing, 2006.

Loretto, Penny. "Penny's Top Internship Sites for 2012: How to Find an Internship." About.com. Retrieved February 2012 (http://internships.about.com/od/internsites/tp/internsites.htm).

Michael, Paul. "How to Answer 23 of the Most Common Interview Questions." *WiseBread*, October 4, 2007. Retrieved March 2012 (http://www.wisebread.com/how-to-answer-23-of-the-most-common-interview-questions).

Paige, Joy. *Cool Careers Without College for People Who Love to Build Things* (Cool Careers Without College). New York, NY: Rosen Publishing, 2002.

Penn Foster Workforce Development. "Carpenter Apprentice Training." Retrieved March 6, 2012 (http://www.workforcedevelopment.com/construction/carpenter.html).

Reeves, Diane Lindsey, and Gail Karlitz. *Career Ideas for Teens in Architecture and Construction*. New York, NY: Ferguson Publishing, 2005.

Sandomir, Richard. "A Distinctive Façade Is Recreated at New Yankee Stadium." *New York Times*, April 14, 2009. Retrieved February 2012 (http://www.nytimes.com/2009/04/15/sports/baseball/15facade.html?adxnnl=1&adxnnlx=1328637834-3seGReRfSsEGCf2/hE+Qqw).

Sohn, Emily. "Hubble Trouble Doubled." *Science News for Kids*, October 28, 2008. Retrieved March 2012 (http://www.sciencenewsforkids.org/2008/10/hubble-trouble-doubled-2).

Tishman Construction. "Internship Program." Retrieved March 2012 (http://www.tishmanconstruction.com/index.php?q=careers/internship).

# Index

## About the Author

Laura La Bella began her writing career as an intern for a business journal. Since then, she has written more than twenty books and works as a full-time writer and editor. She lives in Rochester, New York, with her husband and son.

## Photo Credits

Cover, pp. 22, 65 Bloomberg/Getty Images; pp. 4 Monty Rakusen/Cultura/Getty Images; p. 7 © Santa Rosa Press Democrat/ZUMA Press; pp. 10–11 © Doug Duran/Contra Costa Times/ZUMA Press; pp. 14–15, 48–49, 54–55 © AP Images; pp. 18–19 Kevork Djansezian/Getty Images; pp. 24–25, 36, 39 Justin Sullivan/Getty Images; p. 28 © iStockphoto.com/George Peters; pp. 30–31 MCT/Getty Images; p. 32 © Steve Skjold/The Image Works; p. 35 © Michael J. Doolittle/The Image Works; p. 41 Steven Errico/Digital Vision/Getty Images; pp. 42–43 © Ellen Senisi/The Image Works; p. 46 Nicholas Cope/Stone/Getty Images; p. 56 Joel Davis/The Oregonian/Landov; pp. 58–59 © Syracuse Newspapers/D Lassman/The Image Works; pp. 62–63 Chicago Tribune/McClatchy-Tribune/Getty Images.

Designer: Mike Moy; Photo Researcher: Amy Feinberg